"Human race will prevail…hehe!!!".

"This book is dedicated to all my brothers and sisters ,wherever we are,whenever we are,we are always together..It´s all for you,by You, with Love….hehe!!!".

PROLOGUE-PRE-LOGUS:
 "I would be calm,very calm waiting the end of the world only with a magnetophone recorder playing the same lethany with William Burroughs voice reading yellow pages from any country of the world.Thats the paradise (Must) be" .
To expose the "show"…What is happening,who watchs,who plays on it,who coordinates,who destroys it…hehe!!!":
Today we are mere spectators are there just bursting the "play" completely unaware that we are watching is real.
The Alien actors are real,the story-line played is real,the Alien stage with their alien worlds,their Alien Ships and Cosmic events are real.
Others know what we are watchins is real,and there´s nothing they can do about it,thet cannot change the "play",and certainly the cannot go to another "theatre". And..They cannot go to another theatre where the play is more of their liking,the only thing they can do is watch!.Others like you,me,are spectators taking notes about what is in front of them ,about the play,the actors,setting and props.We are learning about ans writting about everything we see,so that one day when we have an opportunity to change the story-line,we are prepared to do so…But beware ,among the spetators there are many that are not paying attention to the play,they are only watching the others spectators around them,specially the ones taking notes and those spectators look just like you and me,the can be seated right beside you, sometimes they even are the one´s that gave you the ticket for the play,and sometimes they are your own parents. "The Black Mountain Dominions Prologue.

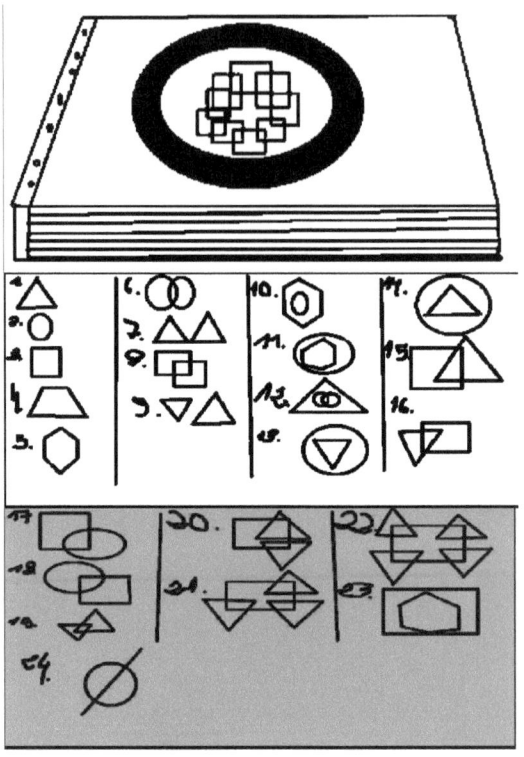

E-psi-Code one : "The black mountain dominions".:

"A litle boy, fattie and smiller,walked in ,pathetically ,he jump,better, into the rocks , from that plateau,that frozen plateau.Here and there pieces in the sky surrounding him to give another reason ,another one, to the snow storms,and give to that moment and place a sacred aspect,silence from snow blanket, as white slices,at those disgusting times…Simple gigant organic robots were on batle mode on desert beaches at the distant red planet,and at once first explosions blow up..."

From Exocosmobiology,No-Renounce Treaty,Plarraph-Dialetical Plaph-electronical diamantine and X°.Single pages from new-nueva ola Emperor.

Gets enter Leonard Dawe into a deep picture scroll ,a voice sounds:

-Wellcome Leonard,to the Black Mountain Dominions,very very high besides Black Crow,wellcome..We are perfectioning dimensions, human being is much more now, and opening in many planes and dimensions,free.We are wining to inter-dimensional parasites,reptoids ans others.They have come to our planet,this planet is a mortal tramp.After 24 years fighting with them,now is the turn for us to attack and beat…hehe!!!.

We´ve come back with ours Galactic Federation troops,we are at the peak moment of war phase,no time to rest:With us are Andromedians,as Insekt troops,helping humans,to Epsylonains,Arcturians, and rebel troops from Pleyadians, the most ar at enemy field.I have the dominion of all universal troops,once again, and thats a very hard responsability from all times on me.But i dont have any doubt,i know what to do…hehe!!!.

This war is a dimensional war too,thats everybody knows,form a way or another we all know.A war with many frontlines.

-Today i´ve dreamed with a katana,that it was my own soul…thats what they told me.

Today,178 Cylon from Second Epsylon Area (2nd Year) 21/02/2015 gregorian calendar,after mnay months,almost 9 form a complet servie to the planet writting and sharing this now i´ve been deleted form all those services an can do what i really was born to do,it says, be an strategic mind.Writting is an exhaust exercise for me,i cannot occupy all my time in this,as a writter must do,but i´ve done by some reason,planetary servie, i think.And it´s been really so freely.Now i want to write "The Blue Letters".

NOTE : Confirmed,today Sudan´s President 21/02/2015 greg cal. Has confirmed what we all knew,Islamic State is by direct creation by MOSSAD and CIA,and finaned by them.Now,to share!!!...hehe!!!.:

"-Phodolx Planet,Angel Constelation,listen to them,please take notes,fast!,they cannot watch us or smell us or feel us,but we can do to them…hehe!!!.

-Cold war reminisences?.Argüelles alter-ego says.

-15 Sector?You seem as one of those psicopathicals form the governments.My Alien alter-ego says.

-It happens everywhere,not only the Earth.

-The question is that human exists out from there and we have to invent all these "dimensions 4D-5D" ..to share with them,is the only way they can wake up.

-Everythnig happens in all time and space .

-If all has happened yet,why we must exist?

-We arrive to this planet,that the only important thing.Take this sentence from Epsylon:"You can always choose"…hehe.

-Very good shot!,thank you Valum…hehe!!".

From "Argüelles en Reconstrucción,el Reino de los Ángeles"/Argüelles in Re-construction,The Angeles Realm",Pag 66.

-Jupiter and many planets and planetoids from Solar System at 3D comes from reverse engineered to another extraterrestrial human and not-human civilizations,to help to several purposes.

"Tecnician" on explosions "Burroughs-Barreiros".

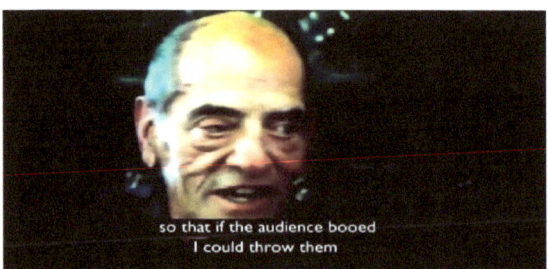

-Buñuel. Jean-Claude Carrière : "When we talk about Buñuel in England or France,it´s about the same,always dues the word "surrealistic".It belongs to a certain marginal way of making films,its not a real classic.But in Spain is exactly the opposite.No one spanish said he is surrealistic,is just spanish.You know, whats seems to us to be strange,rare,bizarre,weird,is perfectly normal in Spain, it belongs to a "strange"spanish tradition".Of course, this a the opinion of a frenchy (whats normal for a french),so we must take care with this,though the structural analise is correct.
-Well done, Leonard,im not more invisible,i can tell you, i´m William Bourroghs,voice said.

Video Youtube : "España es Idiota" Def Con-2.
From Epsylon Galactic Clearing Up Center (E.G.C.U.C)Desde el Centro Reparador de la Galaxia Epsylon (C.R.G.E.):
-Which is the solution,William?
-To come back to Nature, is the only option, because we ara a prt of it, and to her we´ll have to return,and leave squizophrenia we are giving to men ,to live in, a gigant concentration camp so big as the planet,a living MATRIX and is a"crap".Fortunately,neo-conscient beens are taking the main positions in this conflict.Taking for them risk and responsability places.
-U have Faith!.You are going to change everything!.
-"(048)43" : "OUSIROUFOREITOUFORZRI"(X22).

-
"253689#895647*1232":"CHOUFAIBZRISIKSEITNAINOUEITNAINFAIBSIKSFORSEBEN OUGUANCHOUZRICHOU"(X22).

Epsi-Code Two : "Hidden Agenda" :

-This is the big deception about New World Order,about global fascism :And this is the biggest weapon among it: Intergalactic Security Minimum (I.S.M.)
We represent "Intergalactic Security Minimum",a concept born into our DNA.When at 1991 NWO tried destroy this concept with a new weapon they were training many years stopping continuum flux and creating a disruption into space-time,an "incertitude","the" incertitude by they were getting into this reality and blowing up everything they could,and everywhere,at once they realized,then, it was the only way to forget the Intergalactical Security Minimum was to look for and fisically destroy who incarnate in every galaxy or group of stars.Since that moment a crazy hunting begun to find and destroy these beens ,they incarnate into their bodies these concepts, and the ISM itself,into their genes.Most part of them didnt know it,to keep under protection whole universe .But now we all know this concept, we discover conspirancy, beetwen our own dreams, a trap behind a trap,and we´ll see the show "behind the curtains",they dont have any weapon,they only need to make us believe psicologically ISM doesnt exists at all and enter ,get ready and enter into our galactical,psiquical and neural system equipment. And they can do this,but when the "Guardians" wake up and realize they wear this ISM galactical make up,internal make up,or they incarnate it, whole mechanism get inwards already and attacks them!..hehe!!!.And in this all religions have got involved,thats the reason because religiuos phenomenon brought up infront our faces since 1991.They colaborate with reptoids in their psicological weapon,in fact,religions were put in action by this way to create, justify, and teach us ISM´s disruption as a normal cosmological function, and preserve this disruption.
-eeeeeeeeeeeeeeeeeeeeeeeeeeeeeeeeeeeeeeeeeeeeeeeeeeeeeeeeeeeeeeeeeeeeeeeeeeeeeeee!!!.
-Yeah!! Wow!!! Well done, William.!!!
-In not mixtured with drunked Ketamine people!,i dont maka any deception to my brodas!!Im not selling my kingdoms to Aliens,by drunking endless ketamine hignways!!!
-Whole codes of mine is same making them by one direction or another, when they were injected into Alien matrix, they are fractals of the total result.
- Please, tell me,an advice for war!.
-Dont Forget to breathe!
-We are going to Take off everything.
-This is a a Mighty sentence….hehe!!.

I see the desert in your eyes.

POEM TO MY WIFE= POEMA PARA MI ESPOSA =
"Princess of The One Thousand Worlds,
/Princesa de los 1000 mundos,
I want drawing new colors
On your skin.
/ Quiero dibujar nuevos colores sobre tu piel,
I want paint new colors
On your skin. /
Quiero pintar nuevos colores sobre tu piel,
And get into one only of your Kingdoms…
/E introducirme en uno solo de tus Reinos…"

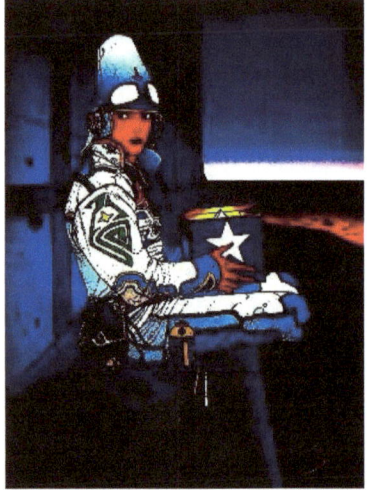

"The statues fallen down when The Big Door´s open"
"Omega",Morente and Lagartija Nick.
Based on Federico García lorca texts.

-I´ve not done the Masterwork yet.I must do it . Sharing together infinite sadness from Granada by lost Al-Andalus ,making it rise up once again,form the depts of thousand miles from Sierra Nevada that feed in the inner sources of the hell and they rise as gigants to thousand constelations sky with Alhambra structure,gigants´land,i must do it yet,and crown this land getting hidden treasures out,and healing bleeding wound,healing biggest wound;and healing whole universe"

I want to acknowledge gauchos´people,with an open wound too, gigantic hurt,getting andalusie Granada treasures out, universals,they are universal!,Leonard said.

"Insekts,almost now can colonise planets sowing their spores to space" from Starship Troopers Film..

-.It seems as i´ve connected once again with insekts colum,well stablished coordenates at Epsylon Eridani, human colony,insekts going on "giving" me their defense and attack weapons,and in our own benefit,is possible. An inner desafection in reptilian forces, insekts are part of,constantly are helping us,of course, with security mesures, to the fact "per se",to be enough human,not too long from humans, a mistake we´ve done twenty years ago.Maybe we need this mistake…hehe!!!.

Epsi-Code Three : Empty Planet:

-Greys= Pure (Organic) Robots= Only Organic Robots.
-Inner sun, following our own hypothesis,Inner Sun,in the last few books we´ve been drawing the keys to realize just NOW the truth without any fear.Its howl and ocuppied by Epsylonanian troops, allies to The Resistance and others too, as our own planet,with test videos.We wacht in the beggining of our "tesis",as when we said in "Las Alas de la Libélula-Presciencia Insekto"("The wings´dragonfly-Insekt Prescience") on theses questions and Insekt-Epsylonanian-Empire...hehe!!!.
-Very cool,Leonard!!!.
1st DAY :"Insekt Transport Spacecrafts"

Epsi-Code Four: "This you can trust!":

-William,i want to present you "The Council of Five":

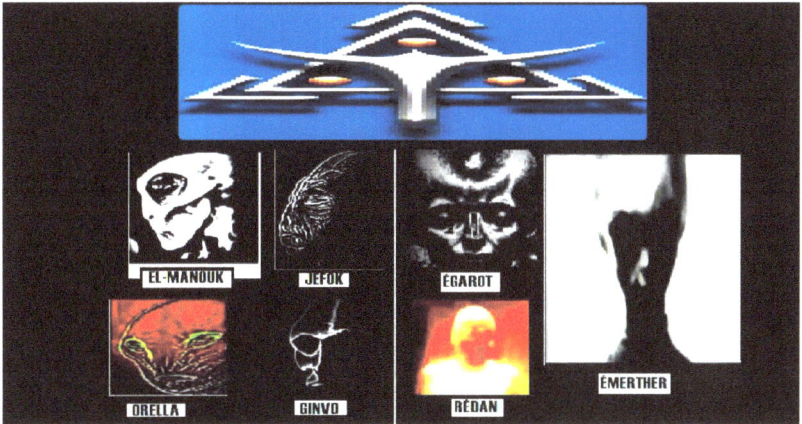

-But,they are seven!!!

-Well, its only the name, you know...hehe!!!.

-Its like "The Five" i read when i was young,five boys and a girl,"jorgina"...hehe!!!.

-Yeah! I remember well, butter,bread,teas..They were always eating!!!.

-Yeah!!!.But can you tell me why you present me "The Council Of Five" ,Whats is that?

-I think is better they present you personally,one by one!!...hehe!!!.They presented to me tonight,Jefok and Emerther.Jefok said they came from "nava Lenticulara" in orbit around the earth at 5th dimension, built by Sirians and they have received my call,then he drawns onto my front in my head this symbol:

..and Emerther, two people of them were backwards him while he does this,saying:

"-You will be a messenger of us since this right moment,an intermediary."I felt very comfortable and soft while they were with me, very happy,and they were very sensitive,slow and gentle to me...hehe!!!.Coincidance? I dont think so!!!...hehe!!!.

-I see, Thank you,please present them to me!!!.I want you acomplished your mission,leonard!!.i love this extract from "Conan,The barbarian":Father´s conan speaks conan´s child:"...No one,no one in this world can you trust, not man , not woman , not beasts....This you can trust!":

"This you can trust!".

Epsi-Code Five: "There´s none here,they´re gone!":

-Arbitrary execution political parties decissions , I am the Party=I Am The State=I Am everything, at USSR´s Josef Stalin´s regime. Leonard said, something angried him vissibly with.

-More or less ,something like that ...

A new voice takes place infront of them :

-To better understand whats happening in the world,right now, we must take a global vission,a wider vission,as a residual and local part of total human race.Well ,better said, our world stopped at its own evolution because a regression , explicit and implicit regression done by these planet political rulers as human evolution gets at its lowest levels, as a total *per se*, not getting the higgest levels and all its potentials,tecnological,spiritual..etc..And it has been done several thousands of years ago, i told you, million of years.When terrestrial get knowing others´humans existence out own terrestrial orbit,other colonies spread out. At all our solar system,and beyond.Human colonies has been in contact with us more recently, telepathically at once and information flux is constant.Many "guru" cientifics and most part of politicals in our world knows about these colonies existence and keep with no word because want to preserve control and their privileges here,at Earth.Since thousand of years humans exist at Mars,and without no contact with us, they are perfectly martinas, but humans.In fact we can say with no fear our genesis as a race takes place from human martians they devastate this other world."Terra".From this point of view all is clear,as our planet as not most important for human rae, but a single point in universal conspiracy,where humans colaborators with reptilians and New World Order,by the way.Comunications with humans and aliens at several and diferent sectors at Earth and colonies are always in function.As you can see at Internet or Deep Web to take these sources.,"just a litle bit" because The Resistance as you know not only Human Resistance or Terrestrial Resistance but Universal Resistance,we have could do this very rently,as well.

Everything takes a new dimension in our world, a new direction,we can discover because things happen and are transmitted to us by a certain way.

Because of that we feel things we cant explain them at all,but we konw very deep intuitivelly these transmissions are not from our planet but out form it and they are calling us to join us to them.At colonies human-Alien comunication is a flux and as a part of those conections is resistance´creation ,not only human, but with all pacific aliens and some infiltrate elements from Grey-Reptoid Empire keep on transmitting us ..We´ll be in contact with you later…hehe!!.

-Whats is it? Who is..?...Then, all is a conspiration!.

-To control human been and he couldnt take totality of his mental,spiritual power.

-Human been is the only one race in whole universe can make things, material things with the only help of its mind.

-There exist sinthetic humans?

-Yes,but many of them doesnt know,yet. Bourroughs said very happy.

-And whats about Moon Matrix?

-Thats all by now!!!!...hehe!!!.

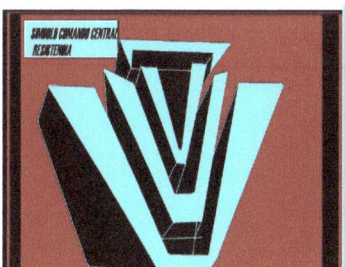

Central Resistance Command Symbol.

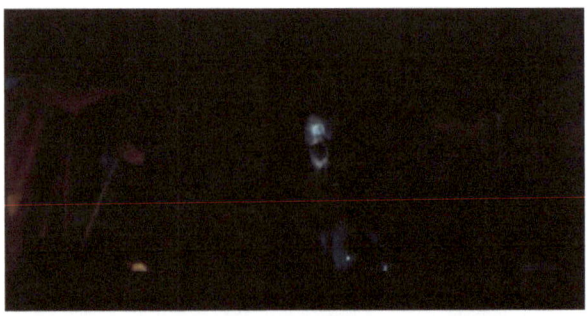

"Almost twenty years of peerless combat!!!No rest,no sleep as other men!".

One of stages used by Francesc I Pope during his visit to Brazil,2013.

Epsi-Code Six: Kuato Lives!:

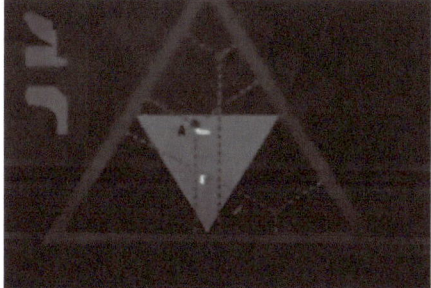

Our Army
So :

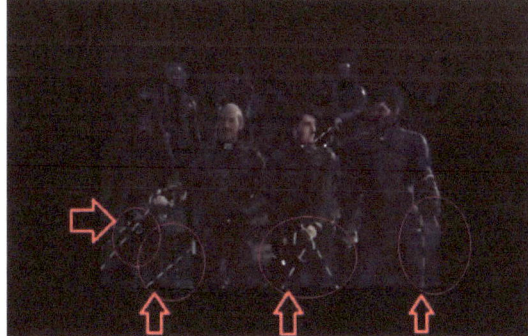

If we had enough kynethic power we could do anything,"Kynethic wepons" and potential Energy,we could have big warmhouses both energies as swimpools and use them with no limits,What would happen then?.

At last film Peter Jakson´s "Hobbit" appears this conflict ver well beetween Pleyadian-Elfs,mayan hobbits and emerthers and arcturian dwarfs against reptilian and all this cosmic garbage,orcs-Trasgs.
Sonic moduls or sonic weapons from Atreides "House" at "Dune" film….hehe!!!.

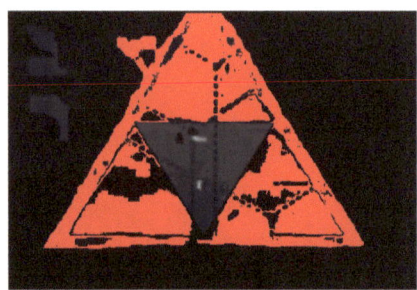

The key from our army.

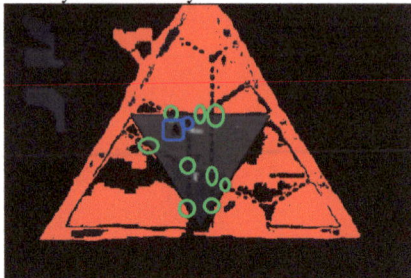

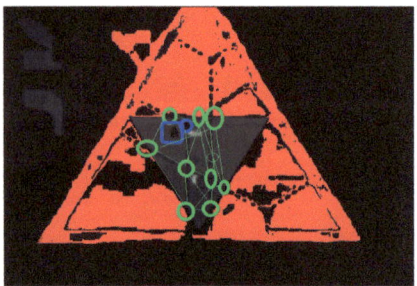

-Symbol used by Trilateral Comission is same as Grey´s.
-Who are the Leverons?.

-Its not the time to make this question. The Navy had the ability to use the time travel technology from about 1970 and developed full operational capability in 1973.

-Did they used that to build the legendary city on Mars?.
-This is the information we have.We have not been able to back it up.It is only memories of Mr. Cameron and Mr. Bielek.I was not involved with that part of it myself.It wasn´t that they built the big city.They found an ancient earlier civilization there at one time and the above ground of it was crumbled back into dust.But they did detect huge undergorund installations which were still making magnetic fields and this sort of thing that they could detect and they realized that there was stiil machinery running undergorund and of course they first went all around Mars and they couldn´t figure out how to get down underground without bringing boring equipment and cutting a tunnel right down in.When Montauk had the working capability it would be nothig to target the other end of the vortex from´83 through to whatever time they wanted to inside Mars itself and this is what they did.And Duncan himself can talk of stuff he saw on Mars.
-What did you see?
He looked to another corner of the room,and said:
-Like there's a gentleman in New York City by the NORAD Michael Morgan who channels an entity known by Yokar who is supposdly a disembodied entity from Atlantis who's up in the fifth dour dominion or the sixth.

Exocosmobiology-Kuato lives…hehe!!!

-Today 06 03 2015 greg cal i´ve had a 'Harkonnen" nightmare typical,as with all their stuff and it really schoked me,a true psiquic-psionic attack,the question is ..from where?.More questions developed now,more than certitudes or answers,iím in a way,keep on a search mode.I´ve beak to my furniture,side by side with my books,and i´ve found this energetic gravity center once again. Plastic that cover up the carpet didn´t leave me to aarive to this energetic planetary center.There comes new attacks from extra low frecuency machinery near my house that disturb my sleeping mode.Could they be genesis of all this? What kind of energies are we facing off? We keep on…hehe!!!.

Answer: Its has all with Mars,and you are in redskywalker geografical zone from mayan sincronary´s Jose Argüelles,here at Rio Grande do Sul.,and today is red skywalker too,and you are living tragedy happened at Mars,in all their details. red .You can wacht "Vampires from Mars" John Carpenter´s film and you´ll have more information.,and you are connected with real human colonies at Mars,transported later from the Earth since thousands of year..It´s strange!!!..hehe!!.Somehow i´ve enter in "resonance" with Mars, with its past-remote-future ans bigs implications for our own world.And woman as dterminant fator in all this interplanetary conspiration from New World Order. and they would de the origins of our own humanity.

-This same information mainstream arrives to people with same frecuency and we vibrate, so, its not new what happen to you, it has an explanation,and its becomes the end of human civilization on Mars, a generalized "vampirization" or "zombification" spread out by "harkonnen" reptilian sectorsto destroy genetically human rae, or trying to shock biological-kynetic sources,a lesson we have to learn about.Thank you…hehe!!!.

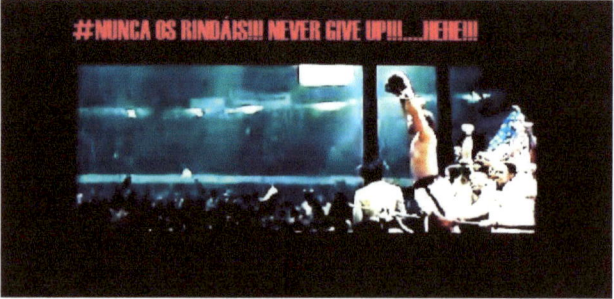

Epsi-Code Seven :" Subliminal Warfare" :

"Ignore Alien Order"- "The Clash :"Rock the Kasbah".

-At Spain there were some years where left was strong,not as a marxist idea, but a concrete way of doing things.There could be out of NWO some several years, a progressive youth power,but it gives to a brutal entrance at NWO at Aznar´s hands.It creates a "culture" that grows up every litle time in many people,maybe as flashbacks,we could do anything,we can do anything.Very far away from the System,the Machine,the canibal capitalism,Alien Empire e*t allie*.

-As Roman Empire times,where cultural system were supersizing another type of system.

-Yes.

-As a regulation of everything based at equilibrium,mixture and open-minded fields.Regulatory culture,as a true democratization of everything, universal acces to culture,progressive cosmic goals, as a new standard that grows up in new-fields all the changing-times;and that´s what we have to do today,from now on.

-A supersizing cosmic culture based at neo-classical ilustrate moods?

-I couldn´t say better.As mental neo-classical ilustrate moods.I read in some place mayans have many contacts with neo-Indian-Greek culture,their time concepts embrace this ilustration mental always-open-minded-proto-culture laying soil of everything, a new *semper viver* philosophical way of doing things.Greece has its roots at neo-cosmic-space civilizations, and all these alien races were in contact with these greece thinkers, giving them laying-openning-minded structures always in increaising new-spin ways,and José Argüelles is best example.

-As pitagorean-greco-roman mathemathical teories into mayan time-tzolkin building. You must forgive what they done to you.

-I know i am very angried with System,right-sides political parties are at my focus,i stabilizied my vision and give to them in their "right" mesure,but i prefere to live under left-sided political parties.

-I think you are at cosmic socialism.

-You are not wrong, but laying at same time at neo-pitagorean-mayan time capsules, we must leave the idea of political solutions,materialism faces mayan way of doing things.We must think in a society ruled by mayan buildings of time, an always-open-minded filosophical way,into collective (not collectivist) moods in the practice.It´s the way you can live without stalinism and without imperialism,both are collectivist,one by "big brother" classical-satanic-tecnological-non-electronical society, another by coercitive machinery,neo-satanic-alien-electronical society,or the machine.Obama has put in contact both and Dark side is complete, Alien Empire is fully complettted.The Resistance has done this work many years ago,we have neo-mental tecnology at fully-function mode.

"We´ve never go so far!"

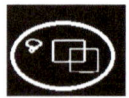

Epsi-Code Eight: "Reverse time Universe" :

-How about some interesting technology spin-offs from the Philadelphia experiment?
-Well, there are a lot of them in use by the CIA and the NSA, as well as other corporate and government agencies. There is a portable unit that can render an individual invisible. The USA is known to use those on a fairly regular basis. There is also a UFO research based covert organization that is bel ieved to have them
- I think a lot of people are aware that the same forces that control the United States today are the same forces that supported the buildup of Nazi Germany and the Soviet Union and arranged for WWII and Vietnam
- In September and October 1990 there was an alien group from some other dimension that was attempting to invade the planet. They took down all the zero-time generators all over the country. The FAA was especially affected. The rogue group was stopped by another species.

For many years, some factions of the Orion group depended on a ring of alien satellites that would sustain life functions. Those were wiped out in November 1990 by the same group.
-So there are positive light forces out there that are seeking to balance these negative activities by the Orion group?
-I am not at liberty to tell you their identity
-Have you watch David Cronenberg´s Scanners?.
-Yes, they explode a persons head to get rid of them with mind powers.
-Since 1947, there have been components of the 6th race incarnating on the planet. The 5th race was the Aryans. The 6th race humans are 100% telepathic - the secret government and the Orion group sees them as a threat. They've been aware of it since 1942.
-In that film, there was a drug that made babies that were 100% telepathic.
- This kind of thing has actually happened. There was a Canadian company that was producing a drug which turned out to do just that. This was between 1946 and 1947. It was removed from the market immediately, although its use continues privately.
- In the movie, the government had a way of using electromagnetics to explode
Sounds.
-Weird!.
-There were eight projects ongoing that also had to do with the development of weaponry against aliens. In 1989 the Orion group discovered this and destroyed the projects.
-Reptilians Winged-Draconians
-And secret projects from 1940: Project called Mindwreaker that would allow paralysis of the mind. The aliens were heavily involved with that project . It produced several neurological weapons, some of which are used on the 8-1 bombre, which also contains a lot of alien technology. At time, various alien species came. There was one group called the K-Group, which was short for the Kondrashkin,'the Kamogol II and Giza Groups,the negative Sirians...
There are buried spacecraft and alien technical archives under the Giza pyramid, They had pale skin that had a slight greenish tint and almost no hair. They looked human, and had to bleach their skin and wear wings.
-No way!!,i can´t understand why they did this.
- Dreamscan came on line- It ceased in 1979.It was another "Project" The goal of the project was to gain the technical ability to enter into an individuals mind in the dream state and cause his death.The project was run by the Secret Government and managed by the NSA. The purpose of the project was to provide for a means of covert assassination. President Carter found out about it and had it stopped. The hardware of both is still intact and in storage.As Project Mindwrecker ,the technology of cloning and the development of synthetic humans and political replacement programs.
-It could change our vission for ever,about everything, if people get to know it.
-Yes,Leonard.Many of them died just to get rid of this mission for entire world,many have been sacrificed,death is not the end.
- What they essentially proved they could do was that they could control a person that they had the "signature" for.A mental message was put out that if anybody heard the messsage they were to call a certain phone number. Over 600 calls came from all over the East coast all the way down to Florid. It works.
-Mental signature?.It sounds tiranic,"Big brother times".Electronical skill control people.What can we do?.
- There is not only an individual signature. There's a racial signature and also a universal signal for the human race. The government has used all three to target specific individuals. They have also done group messages targeted on a specific racial or ethnic group. That's common. In Boston and New York they were doing experiments on "mood control" on the cities.
Transmitters used no longer exist, but the technology does - they can ' t get everyone, because some people are naturally resistant , depending on their level of mental and psychic development. Perhaps 5% of the population do not respond to these signals. If they get 95% coverage, they don't care about that 5%. That's what they have the riot squads and the concentration FEMA camps for. There is no defense unless you can interfere with that signal.It

can be recorded and stored and replicated on a computer. If they can replicate your RNA/DNA pattern they've got you too - for life.

-You mentioned about concentration FEMA camps?

-Yes. They are all over the United StateS.

-You 're suggesting a state of absoulte corruption.

-Absolutely correct. Planned corruption.

-With this kind of technology, why do they need concentration FEMA camps?

-Because there are always people that are resistant.

I would think that they would annihilate them, not lock them up, The first step is the camps, where you can handle them easier, then you can eliminate them en masse. You don't go down the streets shooting everybody dom. Once you're in the camps, they break you down mentally. They're experts at that. We think they are pre-shipping guillotines in there for the people that don't comply.It creates a slave labor force in the prison system, which will be privately owned. States will pay the private prison a fee in order to put their prisoner in there.

- There's a fellow who calls himself Lord Mattreya who says that he is the Christ returned to the planet; that when he links up with the international press he will send a telepathic message to the entire planet in their own language and they will hear it. They also are supposed to receive a visual image.

-So.. they have that technology?

-Yes,they have many years ago.

-Electronic Christ creation.

-No, Electronic God´s creation, they want to made Jehovah reborn and flew into the skies,a kind of electronic mind they want we believe is God,as they tried with NSA global surveillance model,and all that bullying forces against earth´s people.That´s not God.And in this project all governments are delaing with, and all religious leaders, christians, islamics,everybody…But that´s wnot be God.They are selling us a God doesnt exist at all.

-I cant take it seriously,i can´t.

-You must know the information and share with everybody,said Bourroughs taking his shoulder with strong mighty power,you should, you´ll do!!....hehe!!!.

-Ok!,tell me more.Talk about time travels.

-When you travel through time you cannot come back to the exact point of origin. It has to be later than when you left. If you were to come back to the exact point at which you took off you would be at the same point twice in your lifetime and there would be a very serious problem.

-And germans,they has this tecnology.

-Yes.German technology was way ahead of us. If the war had gone on another 30 days the Germans would have won it. They had super weapons in pro- duction which they were ready to use. They were so close to winning the war that Churchill and FDR were really worried about it. If the system shifts, it could shift to a parallel reality where the Germans won the war.

-That's why the government doesn't dare kill either of The Resistance members?

-Right. Because of what we were involved in, it might jeopardize our entire reality system.

-As a big deal with them?

-No, they know that I know why they can't touch us. If they do, they know that there will be real problems in time and space because of it. There are several of us that are holding certain factors in stability. If anything is done to either of us, there could be a rip in hyperspace. The whole reality system will shift.

-And Stealh project?.

-The Stealth has a drive system which is very advanced and allows it to fly in space. The assistant director of NASA admitted that this came straight out of alien technology. He admitted this to the public. Soviet had scalar weaponry.

-Like soliton and tensor fields.

-It is a fact that the unified field theory was completed by Einstein and given to the US Government. They have it and they donst want anybody to know they have it. It was never released publicly in any books. This kind of knowledge is used as a method of control between and for governments. It's unfortunate. The government is supposed to be "for the people and by the people". That is what it says in the Constitution. I wonder when the last time the president read the Constitution?

-And the International Aerospace Alliance?

-Well, its a super secret international organization that is funded by all major governments. It performs research on al iens and al ien technology, coverup operations, and also does espionage. The group is negatively oriented and is considered to have no positive attributes in relation to other humans. It's called the International Aerospace Alliance.

-And what about Philadelphia experiment at Eldridge boat?

-August 6th, 1943, UFOs appeared over the Eldridge for about six days.One of the UFOs was sucked up into hyperspace with the Eldridge and it ended up in an underground facility in Montauk in 1983. It contained a charging device which some aliens made us go back and get for them, as they didn't want humans to have it. We don't know who they were. The Philadelphia Experiment was not an alien operation, as such, but what was the set-up was the date, August 12, 1943, because it had to be locked to the Phoenix project on August 12, 1983. The date was set by alien influence in order to cause a 40-year hole in hyperspace through which large numbers of alien craft could enter this dimension. It worked, but it didn't last long enough to give the aliens the maximum benefit of the scenario.

-Who was in charge of the project at this time?.

-Dr. John Von Neumann and Jack Pruett.And David Hilbert, the mathematician who devised equations for Hilbert Space, which described mutiple space or mu ltiple realities mathematically. These equations for multiple space became very important in the project. Dr. Von Neumann met Hilbert in 1927 and retained a lot of what he had learned. With that, Von Neumann developed other new systems of mathematics. Von Neumann was considered to be one of the most outstanding mathematicians in this century. Some think he was better than Einstein. Another ma thema tician involved was Dr. John Levinson, who was born in 1912. He died in 1976. He published three books on mathematics. There is no other history of Levinson himself anywhere that I can find. Levinson developed the so-called Levinson Time Equations. With all this behind them, the group had all they needed to proceed with the project . Tesla made periodic announcements in the late 1930's and early 1940's about his contact with off planet species. He was in contact with the outside, who agreed that there was a problem with the people. He decided to sabotage the 1942 test in an attempt to stop the project. He de-tuned the equipment so nothing would work, The test failed. Tesla then turned the project over to Von Neumann in March 1942 and left the project.Pruett was concerned about an alien invasion.

-What about "the creature" at Montauk project at 1983?

- They had the subject in the chair think of some creature, and the creature would materialize. They had the individual in the chair think of all the animals at Montauk point charging into town, and that's exactly what happened. They almost had the power to create a being. The problem they had was that what they created only stayed as long as the mind amplifier was on. The power was somewhere be tween gigawatts and terrawa t ts. Tremendous power. The vortex could have a diameter of about five miles. The creature ate people and equipment.This was on August 12,1983. The vortex locked on to the 12 August 1943 test and formed a loop. All this occurred because someone planted the thought in.One can see from the way that thing was maneuvered that there was a higher force of some kind involved in setting the whole thing up.40 years apart coincided between the two experiments: the project known as the Philadelphia experiment , or Project Rainbow, in 1943 and the Montauk project in 1983.Now, the Earth itself has a biorhythm that peaks on a 20 year cycle on dugust 12th. It "just happened" to peak and provided the connecting link through the fields of the Earth for the two experiments to lock up in hyperspace.One of the first things they did was send recruits forward to around 6030 AD. It was always to the same point. Somewhere in an abandoned city where there was a statue of solid gold.Yes. You can also change the present by sending someone into the future. Under certain conditions. The government is using existing time machines to go forward in the Montauk time line. Its linked also with new life form masses over the poles and their relation to yearly outbreaks of flu-like disease, AIDS and Fort Dietrick (NSA),maglev trains and the US underground tunnel network,the missing human genes, electronic life support systems of the Reptilian Humanoids …

-Are you saying that now the present can't be changed because we have established a time-loop through the future and the past?
-Yes. There is a very specific field identity pattern you can detect if you are sensitive enough. You can spot a person who has been through the Montauk program "five miles away".Around 1979 or 1980 it was fully operational.
-Did they send a lot of kids down the time tunnel as i heard?
-Yes, but we don't know what their goals were. They lost a lot of kids. We estimate they took 10,000 American children off the street and brought them to Montauk, New Jersey. The total number of people that they pulled off the streets for the 25 stations was about 250,000.
Suddenly a flash of liht appears beetween me and Bourroghs at my left side twice in two mechanical minutes.
-What?..Why?
-They are watching us, Empire Alien.We must be calm.
-Please keep on..
-Yes, it works effectively in the period from 1969 to 1971. Phoenix One went from 1948 to 1968. The first part of the mind control project was to take an individual and stand them about 250 feet away from the antenna. The SAGE radar had a peak pulse power of .5 MW. The antenna had a gain of 30db. That means an effective radiated power of at least a gigawatt. It was nominally a gigawatt. Can you imagine what that would do to people? I think its amazing these people are still here. It does things like burn out brain functions, create neurological damage, scar lungs from heat, etc. They tried this with a number of people and there were few survivors.
-And so? And where Montauk Project is in all this power structure?
-The Montauk project, was a combination of Wilhelm Reich 's work and the Philadelphia Experiment .Wilhelm Reich's concepts and some of the transmission schemes used from the "radiosonde" project , they found that you could combine the two factors and use them for MIND CONTROL.
-Incredible!!!.
-The result is of orgone energy coming into contact with an enclosed radioactive source. This produces a Deadly Orgone Energy - DOR - which threatens life. Reichs contact with alien species, his discoveries about 1ife energy and cancer, and the mind control applications are some of the reasons why he was killed. This knowledge was combined with other knowledge, some of it alien in nature, and integrated into the work at Montauk and subsequent covert projects to subvert the people of the United States and the world under an Orion based system. Its used by the US Government and Sirians,the Psi-Corps,Montauk and the aliens from the Antares system, the Leverons, the Elohim Group.
-Dimensions involve the 4th and 5th dimensions but your anti-matter universe is also locked in, but it's not locked in that way, it's locked in in the 6th dimensional level and also the 11th. And you get, through this whole, strange arrangment, you get into, through higher order time manipulation,putt get into areas of other universes.The bouncing electron in the atomic shell that's recal for a period of time and then virtual for a period of time. It goes into a reverse time universe.

Epsi-Code Nine:"Figure Out":

-Hold on these key words: Cabala-Quimera=Secret Space Program-Blue Avians-Warden Solar Space Program.
-Why?.
-Just try hold them,ok?.
-But you can tell me why.
-I must keep on giving you information you need .All Strangelet and Toplet phisics boms from Quimera´s group have been erased,they only have plasmatic ones. Galatic Confederation´s Fleet out our solar system implement a barrier through it no space vehicle from Alien Empire´s group

cant scape.Four Reptilian Races waiting for secretly Alien Empire give them an exit beyond this solar system,to Barnard Star as a promise from Alien Empire to first iluminatti. Alien Empire controls some phisical fragments from draconian and reptilian ships hidden at some minor importance asteroids at asteroids belt and Kuiper´s belt.

-Something as militar operations out bounderies.

-Solar system bounderies.This weekend,Light Forces,Pleyades Ships,Sirius Fleet and all Resistance Movement han begun a cleaning operation at all complete solar system.There were by now some Draco Reptilian frag Fleetments,hidding at Barnard Star asteroids belt and 359Wolf Star.All those fragments from ships have totally gone .Quimera controls a BUMP system fortress,with Long Island been its only important acces point at surface.These Quimera´s underground fortress are not form BMSP(Deep underground military bases).All BMSP are cleaned and Cabala cant take them,they are stopped at surface. We need to understand persons inside these space secret programs couldnt get there to Quimera at our solar system,underground our planet and many of them weren´t conscient about their existence.But The Resistance will keep on getting free.

-What needs The Resistance to erase reptilians from Solar System?.

-All Solar System is under surveillance.No one cant enter, no one can exit, and no communicattion can enter or works..at only last 3 weeks,weapons wide distribution,EVERYTHING,massive secret weapons hace been put on line or erased.This has nuclear weapons,SDI,no way...In my personal opinion,as everybody says,it will be before main circunstances going to happen at the Earth.

-It blockes our fundamental communications out solar system.

-Each member of reptilian ships trying to exit has been removed..Many reports about meteors and strange atmospheric phenomenons.

-Meteor reports?

-Not only solid meteors but conflict results,inluding metal trash from a Sirian mothership (more than 300 millions of fast reaction ships) when we wanted to stablish a massive defensive field force ant it was detected..Astronomers wont be able to stablish rational origins at those metal pieces.

-That changes all we know about our world and the solar systems´relationships.It makes our concepts virtually blow up!!.

-Snowden documents are much more we could never imagine,it hasnt been revealed in its totality by Quimera pressures.Reptoids failure is a fact,many have been convicted in our solar system,they cannot go out,their allies cant entry.There´s no communications,and they are very frightened.

-Tell me more, please.

-They were convicted ,millions of them at etheric layer at earth planet after epic combat besides the moon,55 millions sirians motherships versus 122 millions reptilians motherships ,they were totally erased, imploding every ship´s matter after their energy extraction and freadom every pilot to planets and support emplacements,and their desperation makes them to do direct attacks to human beings´ etheric bodies causing thousands of deaths.We were waiting these events to 2016-2017 as "Big Peak" that at these moments are going on.Everybody of us are where we must be,and it will happen what will happen,as contemporany parodoxes.A present cosmic event can be archieved into the past.NASA hás at its hands all information but never could translate present data until its recorded and past events takes light then,and it cant be changed because reptoids cannot remove time lines.

-Argüelles is behind all this, i know it!!!. You´ve been fallen down!!

-No!, I didnt!, I just try to pull muy triggers up,we all must!!

-And watching…

-Yesterday 10 03 2015 cal greg i catched one of them..

-One of what?

-Flyings!!!.

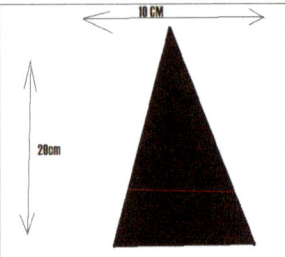

A Flying Been

-I catched one of those damm things!!!.As Castanesa said, "The active side of the infinite".
-I read it, well, as well you did.
-There are many things to say, the mean is..it goes out from me!!!,it scared of me...hehe!!!
-I can´t believe it,really i´m getting insane.And you do,c´mon go to "Andrew´s" and take a beer with friends,we need it.
-No, not now,later i call you,..Congratulations!!!.
-Im afraid of you!!!,Yo are really so...Well ,clap your hands in the air!!!...haha!!!.
-Raise your hands in the air!!!.He said with a very funny mood.
-Now i know why i make all my codes,because <u>from Marz</u> they codify "everything", and thats the reason i put in contact with all this stuff and all their minds and connect with them, or i decodify all codes from Mars,thats what im doing since 2002.Everything in our own Planet is about Mars!.

<u>Epsi-Code Ten: "Conversations with The Lenticular Spacecraft (about Lyrians and what they are going to transport Planet Earth)"</u> :

-I´m going to give you "Granadian tomate sauce".
-What?
-This is from Granada,Spain,vey popular,you mix with rice or spaguetti,as you want:
You put up a litle bit of olive oil in a big pot, you place a big garlic´s head and some more slice´s garlics, then many onion rings and green onion.Five big ripe tomatoes,when they are done and smasehd into the sauce,you place frankfurts and all meat you can (from another big meal,aka. Mexican Pizza or burritos,or quesaditas..).If you want you can add a glass of red wine (spanish or italian..) and let heat up 5 mecanical minutes and oregano´s industrial large quantities in all operations,when it squeezes all water you plae 20 sliced olives and 12 sliced mushrooms *champighon* ,and mixture all very well and place industrial tomato and oregano´s industrial large quantities and when all is almost squzeed is perfet..Enjoy it!!!...hehe!!!.
-I´m sure!!!.But you didn´t invite me to your flat to talk about it.
-No, Maybe this meal opens our brains to new advanced cosmic goals.
-What you mean?.
- Cosmic life and very advanced technology are presented in a way that greatly surpasses the limited materialistic ideology.You know The Lenticular Spacecraft?.
-No,whats that?
-The construction of the lenticular ship, the effort of dozens of extra-terrestrial civilisations, is meant to produce in the coming years a remarkable transformation in the mentality and knowledge of mankind.You understand? New technologies will be sown from this spacecraft directly to Planet Earth,which will create a major leap in the psycho-mental experience of the human being.
-But this is a Spacecraft? Or a advanced base from alien races?

-You can say what you want, but all discoverings at humanity came from brothers form stars.But they didnt say they were,because ar our own discoverings,they only follow us at our own evolution.They don´t help us, their influence were present at Greece, many philosophical ideas from Mediterranean lands are parts of this colaboration,they only appear when we call them,is very different.

-Whats its name, the name of this spacecraft?.

-Mothership Anais.

-Whats its size?

-It´s about half Earth's diameter.Colossal structure being constructed onboard the Mothership and many Races are involved.But Sirians are most important part in this intellectual-technological building.Do you understand? It´s the most collaborative alien-human construction since greek philosophies , doing possible all future techonological-technical developments .Knowledge is the gift for humans form other alien races, we´ve been so involved in this cosmoc conflagration, mnay of them that they want to give us this as a Gift, a present.

-But could it be a "Troyan horse" to completely invade planet earth by some dark forces hidden as "benevolent races?.Remember Sirians B and all deceptions we have discovered.

-You are right,we must be very carefully.It can be a future major warhead from reptilians.The ranswer is that as free will and availability are major factors, it is not always possible to mobilise a huge alliance on demand.Anais was under attack last months, and had serious damages.And many humans colaborate in this war to protect Anais. Over 70 civilisations responded to repel the attack; plasma from the Sun was drawn to help restore the damage to Anais; Anais the Casaul Being was also contacted to provide new data for the repairs and reconstruction, atom by atom, in a highly precise and complex operation

At this same time, the Sirians and Arcturians devised a plan to aid our situation and transformation

-I can understand this, last months i had a very carefully work,letter by letter,book by book, a very discipline artwork and it has its counterpart at Anais´construction.

-Sirians make most part of this hard work, at scientifical and etherical side,this involves the construction of a huge special vessel (3500km x 1200km) that would be used to transmit energies that would facilitate contact between Humans and ETs Physical contact "team" (about 100 million Beings) from the ET side will comprise representatives of the civilisations that formed the rescue Alliance; their "fingerprints" are encoded in the construction of this special vessel, shaped like a lenticular spherical concave lens.This ship will transmit "Emograms", a holographic representation of each ET volunteer, complete with emotional elements, for direct "contact" with specific Human counterparts, being able to adjust in accordance with each Human's particular energetic qualities.All Cosmic bodies (stars, planets, satellites, comets, etc) are Beings in their own right, and therefore have their own consciousness.Therefore, this special vessel that's being constructed onboard Anais, is her "cosmic child", in conjunction with her male counterpart, A-nais; this vessel will have its own "incarnated" casaul Being, who is highly advanced and able to take on this very complex nature of the vessel's requirements.This vessel will take up position between Earth and Moon, on the etheric plane...Many warriors,humans and not humans has give their lives for this project. I think we must remember them now!.

And Leonard and Borroughs put in silence and we were sharing this moments remembering all fallen warriors fighting for freadom in all times and places at all whole universe.

-Then its the moment we were waiting for some many millions years ago,we can dream to get free from all governments,all institutions and create a new world.

-We are one,Leonard,we are one, in all whole universe. Sirians foresaw the Reptilian fleet's attack but allowed it to unfold so that they could turn it into an opportunity to collaborate with other Races to destroy the fleet and nullify the potential threat to the increased frequencies already implemented in the etheric field of Earth; this also enabled The Resistance to end all timeline manipulations by the Reptilians.Its the moment to destroy New World Order.

This operation would not have been possible without aligned magnetic forces of Earth with that of the Sun and Moon; this harmonisation also allows the energies from the Galactic Centre via the Sun to be stabilised and utilised beneficially by Humans.In 10 years, computers and devices will have emotional component - bio-technology.Humans will be "segregated" according to their own choices - positive or negative - and their resulting and aligning experiences
This lenticular craft will also transmit the energies from the future, specifically those of The Event, so that Humans who are vibrationally ready for this state will experience it within
Also serving as "prepatory process" to cushion the energies of "The Event"(2016-2017)
We are also being energetically supported to elevate our own consciousness
Future devices to incorporate piezo-electric capabilities ; also allows Sirians and other alien races to insert hidden/suppresed truthsy Reptilians and Greys onto Internet system.

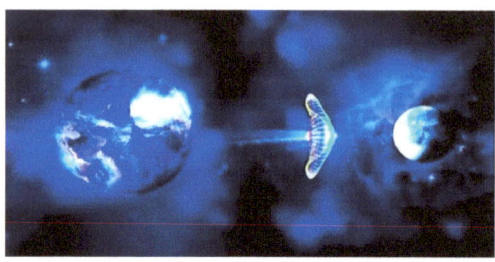

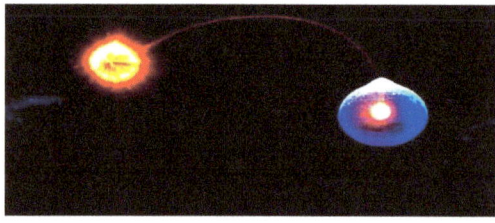

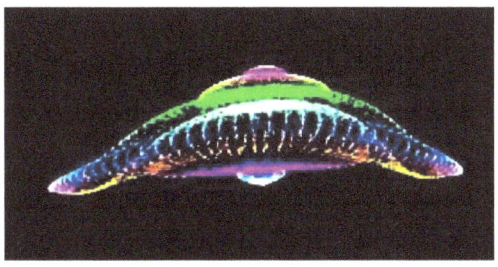

Epsi-Code Eleven : "Re-discovering Sirians A":

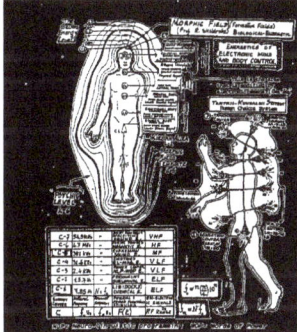

-Sirians make most part of this hard work, at scientifical and etherical side,this involves the construction of a huge special vessel (3500km x 1200km) that would be used to transmit energies that would facilitate contact between Humans and ETs Physical contact "team" (about 100 million Beings) from the ET side will comprise representatives of the civilisations that formed the rescue Alliance; their "fingerprints" are encoded in the construction of this special vessel, shaped like a lenticular spherical concave lens.This ship will transmit "Emograms", a holographic representation of each ET volunteer, complete with emotional elements, for direct "contact" with specific Human counterparts, being able to adjust in accordance with each Human's particular energetic qualities.All Cosmic bodies (stars, planets, satellites, comets, etc) are Beings in their own right, and therefore have their own consciousness.Therefore, this special vessel that's being constructed onboard Anais, is her "cosmic child", in conjunction with her male counterpart, A-nais; this vessel will have its own "incarnated" casaul Being, who is highly advanced and able to take on this very complex nature of the vessel's requirements.This vessel will take up position between Earth and Moon, on the etheric plane...Many warriors,humans and not humans has give their lives for this project. I think we must remember them now!.

And Leonard and Borroughs put in silence and we were sharing this moments remembering all fallen warriors fighting for freadom in all times and places at all whole universe.

-Then its the moment we were waiting for some many millions years ago,we can dream to get free from all governments,all institutions and create a new world.
-We are one,Leonard,we are one, in all whole universe. Sirians foresaw the Reptilian fleet's attack but allowed it to unfold so that they could turn it into an opportunity to collaborate with other Races to destroy the fleet and nullify the potential threat to the increased frequencies already implemented in the etheric field of Earth; this also enabled The Resistance to end all timeline manipulations by the Reptilians.Its the moment to destroy New World Order.

This operation would not have been possible without aligned magnetic forces of Earth with that of the Sun and Moon; this harmonisation also allows the energies from the Galactic Centre via the Sun to be stabilised and utilised beneficially by Humans.In 10 years, computers and devices will have emotional component - bio-technology.Humans will be "segregated" according to their own choices - positive or negative - and their resulting and aligning experiences
This lenticular craft will also transmit the energies from the future, specifically those of The Event, so that Humans who are vibrationally ready for this state will experience it within
Also serving as "prepatory process" to cushion the energies of "The Event"(2016-2017)
We are also being energetically supported to elevate our own consciousness
Future devices to incorporate piezo-electric capabilities ; also allows Sirians and other alien races to insert hidden/suppresed truthsy Reptilians and Greys onto Internet system.

Epsi-Code Twelve: "We follow interviews with human colonies in other planets…":
-Do you think? That we,humans, are living at other worlds?
-Many human colonies have stablished since many millions years ago.
-At our galaxy?
-At our solar system.
Suddenly at command resistance computer one signal was decoded .
-What´s that,Bourroughs?
-We´ve catched a private communication beetween two officials russian army
-Where..?
-Its at Deep web,man,its not difficult to find..I can´t believe it!!,Listen:
"-..from planet to you…got it?
-Yes,they confirm...
-..then,you were born out planet, isn it?
-We´ve got a lot of time here.
-We have not visited the Earth ,yet..Its a litle bit complicated,we can´t travel
-It must be really sad.
-Its sad you cant come here too.
-But..how do you get there?
-We ..didnt travel..We have thousands of years here.
-This is hard to understand.
-There are colonies not only at this planet...we live with aliens here.
-And how do you learn the idiom?
-..thousands the years here..there´s no any kind of problem.
-wow..nobody knows about it,it could change everything!!!.

-Yes, but we cant inform it..thats why we code everything...Humans ..really..they dont come from the earth,they come form here,that´s everybody knows..My parents were born at Mars,and my family too,nobody at my family is terrestrial.

-(a third voice)Everybody talk in Russian there?.

-There are many languages all around the colonies ,and martinas too, there are hundred..There coming humans here..

-Oh,awesome!!!.

-How do you travel there?

-Portals, south planet..There are portals to travel...we dont occupy ships ,we curve time,we use this alien technology...

-Do you ..curve the time?..."

-What you think?

-Bourroughs,i´m shocked.

-Many reports says there are some aliens at Mars called "The Martians",with 6´5 feet height, and green skin,they eat plancton from caverns,deep underground.They are transparent at some kind, but paying attention you can watch them.At future when humans develope their mental powers we never need houses, dresses or techonologies, we´ll connect diretly with our power minds, and we´ll evolve biologically to easier ways, biologically less complex ,simple digestive-intestine tracts

-Yes, evolution,geological-neural evolution, humans will evolve from biology to geology. I think i´ll understand you,we wont need anything, simply existing and becoming transparent beings eating plancton, apparently,because our minds will drive alone through all cosmos ,and our evolution,cultural evolution will be totally diferent from now on.We´ll never need vocal or internal to talk,we´ll develope hight telepathy. We´ll develop instinct-telepathical organs, internal organs ,mind-organs in connection with more evolved being throughout the cosmos.And we´ll be transparents and eat plancton!!!..hehe!!.

-Ok,ok..Its enough for me…I´m hungry.Do you want a mexican pizza?,and after this conversation,William we´ll need to talk many times,there are many thing to clear.Do you want Macadamy or Strawberry icc cream?.

-Pistache!!,always pistache,Leonard!!!.

-It´s time to fight back!....hehe!!!

Leonard turns his head back and tells William Bourroughs:

-..so,we´ll have empty bodies in an...!!!

-And Bourroughs with a happy expression answered:

"En un Planeta Vacio"..an Empty Planet!!!...hehe!!!.

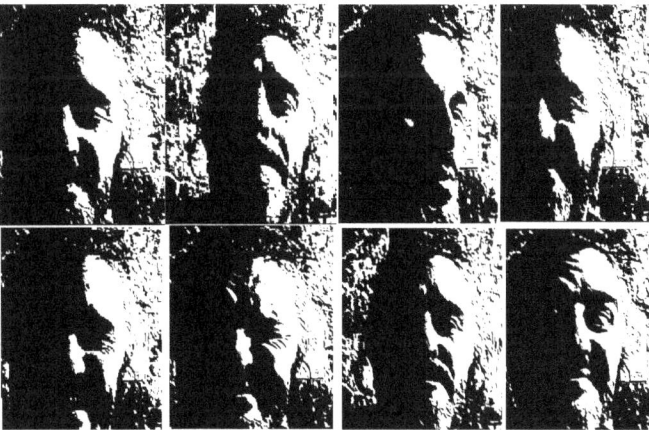

Human race will prevail…hehe!!!".

<u>Epsi-Code Thirteen: "Mother" :</u>

"INSEKT CYCLE : XXV.3.2.4. :
PSICOZOIC ERA : Materialism´reestructuration,and historical also ,that is to say,Marxism,the emancipation movements´big enemy from human being in the begginings of XXI Century .Not only fascisms get into their box-drawers,about slavery, but classic Left,"et Allie",keeping their human being machinery´s idea and social construction happened in the begginings XXIX Century.We got with the rising of communicaton´s new technologies, that is hold processes determination (linked to computer) in favour of new ideas and specialization of bee-worker at Cibernetic Beehive, very similar to a "cibernetics-art-labour"so lefts involved with capitalism that holds it use it at working dimension.We hope this analise is usefull for you.Bibliography about Labour and The Machine,begginings XIX Century.:
1.Machinery History at France (social)=
2.1 1. 1C.B.Allot/C.SCHIMDT =
1.3-PRIMITIVE,PARSONIANS
4.ENGLISH LUDDITE : DOMESTIC SYSTEM.
5. STEPHEN MARGLIN.
6.MECHANIZATION: POLITICAL FUNCTION LABOUR´S DIVISION.
7.MACHINERY POLITICAL FUNCTION.This is for capital cummulative processes.
8.Against worker´s automatization on productive process. : in favour to capitalism,its prescribed work nature or quantities to produce.
Please let me tell you,i´ve read one of Subcomandante Marcos´ Speeches and i´ve translated it by my way:
"Against System Manifesto: To the people of all Planet; to all peoples and governments all around the world:
"Brothers, we´ve born into night,we live in, we´ll die in it,but light will be tomorow for the most,for all of those that cries night today,to all those day is denied,light to everybody, everything for everybody, our fight is to open us to them to listen us, and bad governments dictate greeding and shout with cannons their ears, our fight is about a just work and salary and bad governments smash bodies and shame,our fight is for life and bad governments adress us war and destruction :
Ceiling,land,jobs,food,health,education,independence,democracy,freadom,these were our demands in millions years´ wide night,these are our demmands today".
EPILOGUE : At this land, America needs todat more than ever many riots,fire´s revolution is alive yet,to all deads and fallen to defend the right to Land.
NOTE: God is a bricklayer form Curitiba (Brasil) and is called Carlos Oliveira."
From "Las Alas de la Libélula-Presciencia Insekto"[The Dragonfly wings-Insekt Presciencie] pag.336-337.
-What are you reading William?.Leonard said entering into the flat ,keeping their keys into a small cup,closer the entrance as begging for something else than the fresh air at Autumm night.It has a litle bit neoclassical furniture with bigs columns, very "iluminnatti",and a big anonymous mask as floating into a red sea, the wall entrance.
-Oh! Nothing,a book about´s Psicozoic Era.
-Ah! "The Dragonfly Wings-Insekt Presciencie",very good,i love it!!.Strange and prophetic.
-Well, i like its conception about zapatism and Subcomandante Marcos,of course!.
-"We were born into night!..."hehe!!!.I´m reading this,look, Leonard saig throwing a book : "The Epsylon´s Defenders".
-What´s that?.I dont want to read now.
-At this books talks about many things, cosmic beings, and says that jesus comes froma Mothership from Epsylon Eridani, a planet,and he was part of the crew with Muhammad (T.P.B.A.O.H),Homerus,Quetzaoatls,and Buda.

-And?
-And? This an change our vision for religions, all religions,we are in a distant planet,we were visited by many alien civilizations at the other side of the galxay and human civilizations at all,and all we know is by someone who knews before us.Thay wanted we learnt about God´s concept, and why all these learnings? Why dont they give us certain technologies to change our enviromental lanscapes or our way of living?.No! .They wanted we learned about God,by future times, OUR TIMES!.Damm! William, we are in a line where everything can be scoped from another angle, another vission!, Its a real shift in human evolution!.
-What it says?
-Many things.This book describes some superior beings form Tau Ceti (Tollan) travelling ,not in a space ship, but a time ship.At the other book "The Epsylon´s Defenders" is said tau ceti and Epsylon the same,same human lineage.So on ship was travelling through dimensions to third dimension ,ther was a problem with it,and it crasehd ot disolved.Look here at "The Epsylon´s Defenders":"Cetians or Tau-Cetians : A human race similar as "mediterraneans" or "southamericans" very similars to caucasic humans but some diferenes beetween them,as tied ears, perfect nose,higher body density as their height (5-6 feet tall) and a classical roman´s cutshair, Tau-Cetians and Epsylonanians are extra-terrestrial humans allies with Pleyadians and they are a part of a bigger Federative Alliance with Vegans,Ummites ansd others,called "The Resistance" has arrived to Earth.Tau-Cetians and Pleyadian´s alliance victimised by Grey predators in in the whole wish to reate a ommon defense agianst their mutual Reptilian Nemesis.Tau Cetians inhabitants are linked ina war agains "Lizzies" a draconian-reptilian species at 1920.By some contactees there exist the "Norcans" in a planet around Tau Ceti, they are blonde and migrate to Earth 14,000 years ago. Tau Ceti is one of most populars systems by some modern sicentist have listened radio signals very clears from there (see "Contact" film). Epsylon Eridani,the main star,and the planet as well, knowledge source that emanates many art forms and information to Earth,including my books,Of course!!...hehe!!!.We have created "Epsylon´s Galaxy Modular/Restorer Center" here at Earth.In last 15 years i´ve been involved by Epsilon Eridani´s civilization by some way,and it inspires me discoverings as : Cronological system known as *Insektonotronix*, and Time Counting System known as *343 Sfere´s Cities*,*Mamouths Sincrotiming Calendar* ,*Cylons´System, Siders Cylons*,*Orbital Cylons*,the beings "known" as *The Unknown*,*Epsylon Lunar Cylon Force Field (E.L.C.F.F.)*, *Time Photons*,*Patanjali´s Syncronatiming System*,*Intraterrestrial Synchronatiming System*,*Epsylon´s Warriors*,*The Commander Says*.All this is higly described at next books:"The Queen´s Beehive"("La Colmena de la Reina"),"The Dragonfly´s wings-Insekt Presciencie" ("Las Alas de la Libélula-Presciencia Insekto"), "#Epsylon"("#Epsylon"),"A Ciberwarriors´Tactice Compendium" ("Un Tratado de Táctica para Ciberguerreros"),"The 13 days that shocked the World" ("Los 13 días que cambiaron el Mundo")","The 377´s Law"("La Ley del 377"), "The Jaques de Molay´s Hand-written Manuscript"("El Manuscrito de Jacques de Molay"),"The Eyhezd-Lya,the Seal´s Opening" ("La Eyhez-lya,la Apertura del Sello"),"The Octogons´Book-The Intraterrestrials Book" ("El Libro del Octógono-El Libro de los Intraterrestres"),"The Book of the War"("El Libro de la Guerra"), "The Lost Children"(Los Niños Perdidos) ,"Re-building Argüelles,The Angel´s Kingdom"("Argüelles en reconstrucción, El Reino de los Ángeles"), "Insekt Awakening´69"(Amanecerinsecto´69"),"The Insekt Tikrazy,XXIX Century and The Neural Geology"("La Tikrazía Insekto,el Siglo XXIX y la Geología Neural")..Books you can find at Internet o buying them at Amazon.com o free distribution at PDF´s format…hehe!!."
-Wow!!!. Its a new highly new way to look for new knowledge branches.
-At British Antartica there are two giant about 65´5 feet long hairy Rinos buried into the ice,at 3 feet underground-underice…hehe!!!.
-What?
-They can go there and they´ll find them.
-But,why do they want to kill me?.
"There exists an ancient stone´s ring in a place out of space and time made by the beating of all those who protects you"
-Want to protect me? Where does come this thought?.I feel so lost,as i´m not more of my own.

-That´s beause you are not property of yourself no more. You are property of humanity ,Leonard ,that´s what happens to those who get to this ring,to its knowledge,at its existence´knowledge,and protect it with their own hearts. You are beggining to walk a sacred path that will take you away very very far away,at the deepest of human feelings,to the sacred places ,always common places ,there´s something has awaken,you´ll must find it ,Leonard,to look for it,and in this quest you´ll find to yourself .This quest is for the whole humanity,and you has begun today.
-But,Who am I?.
-Ask your mother.She knows.

Insekt Queen answers: Insekt Queen Awakened!!!.

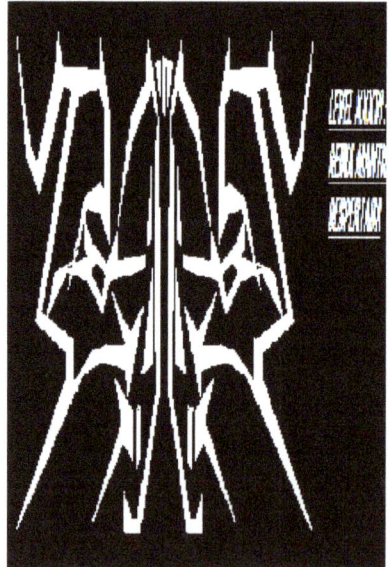